CREATURE FEATURES: CLASSIFY ANIMALS!

IT'S AN AMPHIBIAN!

BY NATALIE HUMPHREY

Please visit our website, www.garethstevens.com. For a free color catalog of all our high-quality books, call toll free 1-800-542-2595 or fax 1-877-542-2596.

Cataloging-in-Publication Data
Names: Humphrey, Natalie.
Title: It's an amphibian! / Natalie Humphrey.
Description: Buffalo, NY : Gareth Stevens Publishing, 2025. | Series: Creature features: classify animals! | Includes glossary and index.
Identifiers: ISBN 9781482466928 (pbk.) | ISBN 9781482466935 (library bound) | ISBN 9781482466942 (ebook)
Subjects: LCSH: Amphibians–Juvenile literature.
Classification: LCC QL644.2 H848 2025 | DDC 597.8–dc23

Published in 2025 by
Gareth Stevens Publishing
2544 Clinton Street
Buffalo, NY 14224

Designer: Andrea Davison-Bartolotta
Editor: Natalie Humphrey

Photo credits: Cover axolotlowner/Shutterstock.com; p. 5 Abhishek Raviya/Shutterstock.com; p. 7 (top right) Ken Griffiths/Shutterstock.com; p. 7 (top left) Tremor Photography/Shutterstock.com; p. 7 (bottom) WildWoodMan/Shutterstock.com; p. 9 Charles Flachs/Shutterstock.com; p. 11 (bottom) Pusteflower9024/Shutterstock.com; p. 11 (top) Jay Ondreicka/Shutterstock.com; p. 13 (inset) Dan Olsen/Shutterstock.com; p. 13 (main) Lillian Tveit/Shutterstock.com; p. 15 yamaoyaji/Shutterstock.com; p. 17 (inset) Kurit afshen/Shutterstock.com; p. 17 (main) Edvard Mizsei/Shutterstock.com; p. 19 (top) W. de Vries/Shutterstock.com; p. 19 (bottom) Iva Dimova/Shutterstock.com; p. 21 TAMER YILMAZ/Shutterstock.com.

Printed in the United States of America

CPSIA compliance information: Batch #CS25GS: For further information contact Gareth Stevens, New York, New York at 1-800-542-2595.

CONTENTS

Boldface words appear in the glossary.

Is That an Amphibian?

Amphibians are animals that spend part of their life in the water and part on land. They are usually found around freshwater ponds, lakes, and streams. Some swim in the water, others hop, and some hide under rocks.

Kinds of Amphibians

There are more than 6,000 different species, or kinds, of amphibians found in the world. There are three groups of amphibians: frogs and toads, salamanders and newts, and caecilians. Caecilians are strange amphibians without legs that look more like worms than other amphibians!

SALAMANDER
FROG
CAECILIAN

What Makes an Amphibian?

Amphibians are **cold-blooded** animals. All amphibians have a **backbone**. Most amphibians have smooth, slimy skin. Many amphibians have four legs. Some have **webbing** between their toes to help them swim. Salamanders and newts have tails.

Slimy Bodies

An amphibian's slimy body might seem gross, but this slime is very important. Amphibian slime is called mucus. Amphibians use the mucus on their skin to help them breathe. It keeps amphibians wet so **oxygen** can pass through their skin.

Amphibian Eggs

Amphibians lay eggs in water. Mother amphibians lay many eggs at the same time. Some amphibian mothers leave their eggs behind. Other mothers, such as the Surinam toad, carry their eggs to keep them safe.

SURINAM
TOAD
FROG EGGS

Larvae

After a few weeks, the eggs **hatch**. Amphibian young are called larvae. They have tails. Most of the larvae breathe using **gills** and can't leave the water. Some larvae eat only plants. Others eat bugs or other larvae!

Tadpoles

Almost all amphibians go through **metamorphosis** before becoming adults. Most baby frog and toad larvae are called tadpoles. As they grow, they lose their tails and start growing legs. They lose their gills and grow **lungs** to breathe with.

TADPOLES

STAGES OF A FROG

Baby Salamanders

Salamander babies breathe with gills. Some salamanders lose their gills and grow lungs as they get older. Many lose their gills and only breathe through their skin. Others, such as axolotls, breathe through gills their whole lives.

SALAMANDER
LARVAE
AXOLOTL

Eating Meat

Adult amphibians are carnivores. This means they eat meat. Amphibians eat any meat that fits in their mouths! They mostly eat bugs, but bigger amphibians eat bigger food. Some large frogs will even eat other frogs and birds!

GLOSSARY

backbone: The spine, or small bones found in the middle of the back of an animal that give its body support.

cold-blooded: Having a body temperature that's the same as the temperature of the surroundings.

gill: The body part that ocean animals such as fish use to breathe in water.

hatch: To break open or come out of.

lungs: The parts of an animal that take in air when it breathes.

metamorphosis: The process of change that an animal goes through during its life to become an adult.

oxygen: A colorless, odorless gas that many animals, including people, need to breathe.

webbing: The skin connecting two parts of an animal's body.

FOR MORE INFORMATION

BOOKS

Brink, Tracy Vonder. *Amphibians.* St. Catherines, Ontario: Crabtree Publishing, 2023.

Maloney, Brenna. *Amphibians.* New York, NY: Children's Press, 2023.

WEBSITES

Britannica Kids: Amphibian
kids.britannica.com/kids/article/amphibian/352745
Learn more about what makes amphibians special.

National Geographic Kids: Amphibians
kids.nationalgeographic.com/animals/amphibians
Discover more amphibians found around the world.

Publisher's note to educators and parents: Our editors have carefully reviewed these websites to ensure that they are suitable for students. Many websites change frequently, however, and we cannot guarantee that a site's future contents will continue to meet our high standards of quality and educational value. Be advised that students should be closely supervised whenever they access the internet.

INDEX